MW01152815

ON AN OCEAN JOURNEY

Animals in Motion through the Seas

ELIZABETH SHREEVE

ART BY
RAY TROLL

little bigfoot
an imprint of sasquatch books
seattle, wa

Surging, whirling . . .
drifting, swirling.

The ocean wraps our
world in wonder,

from rocky coast to
deep down under.

Come, let's go—

on an ocean journey.

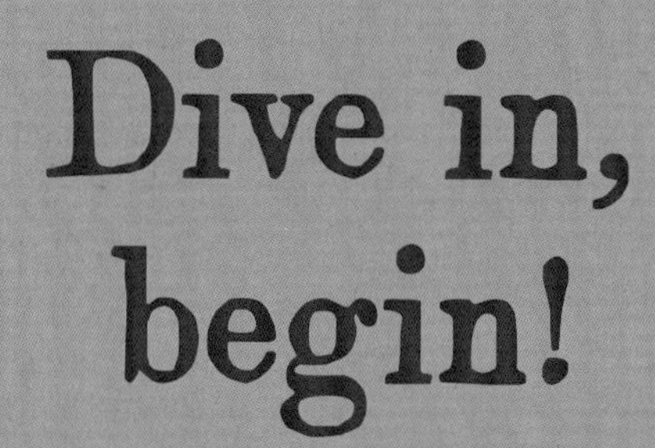

Dive in,
begin!

Here at the shore.

Surf is crashing . . .
foaming, splashing.

Urchins wiggle purple spines.

Fish and seaweed twist and twine.

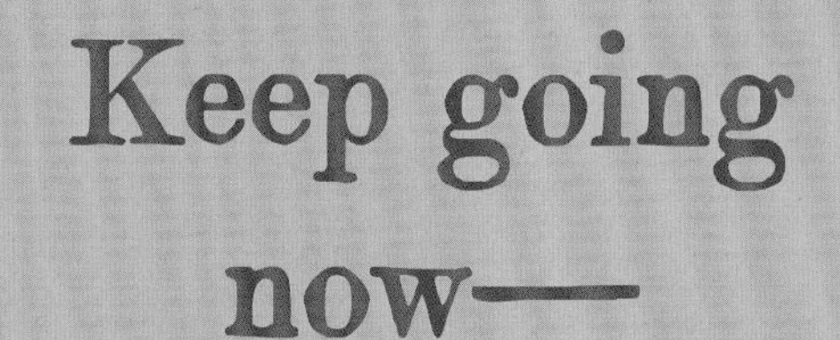

Keep going
now—

leave

land

behind.

Tides are whooshing . . .
sweeping, swooshing.

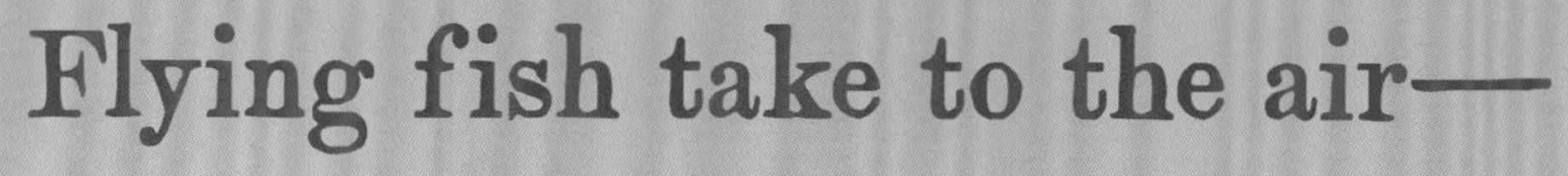

Flying fish take to the air—

frigatebirds swoop everywhere.

Moonfish rise from the depths at night . . .

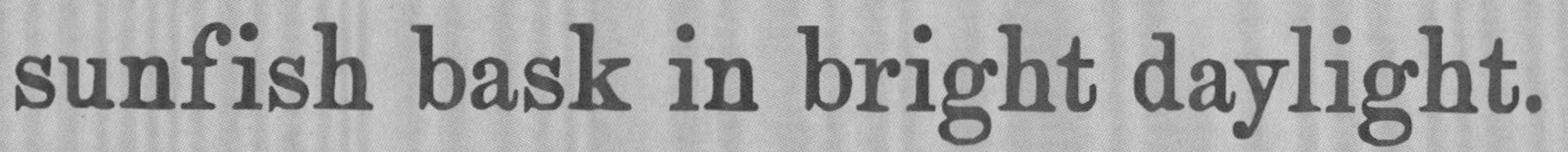

sunfish bask in bright daylight.

Out farther still,
where seabirds roam.

Trade winds blowing . . .
. . . currents flowing.

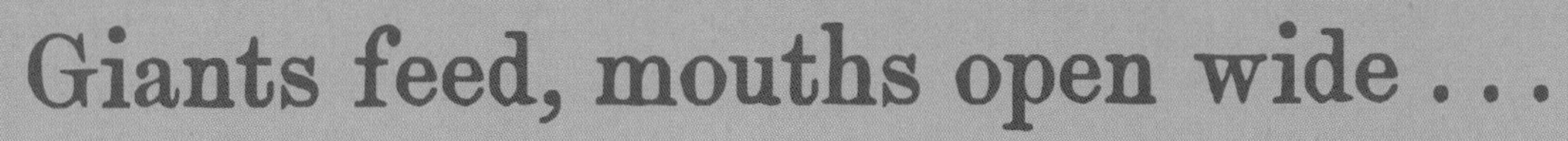

Giants feed, mouths open wide . . .

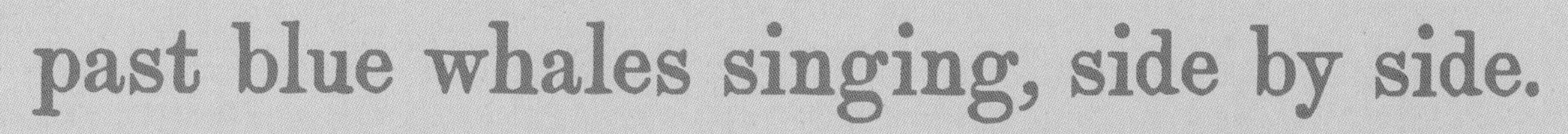

past blue whales singing, side by side.

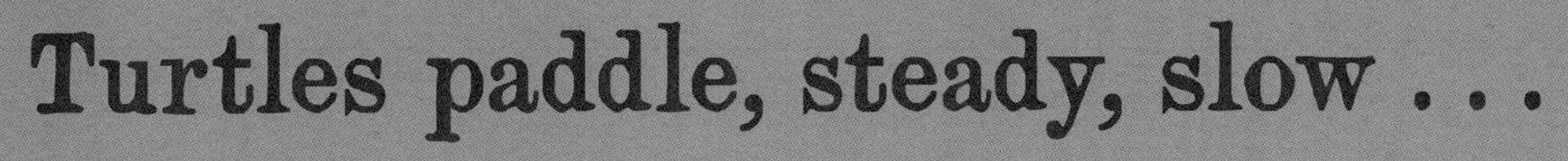

Turtles paddle, steady, slow . . .

gobbling jellies as they go.

Faster are the sharks and dolphins—

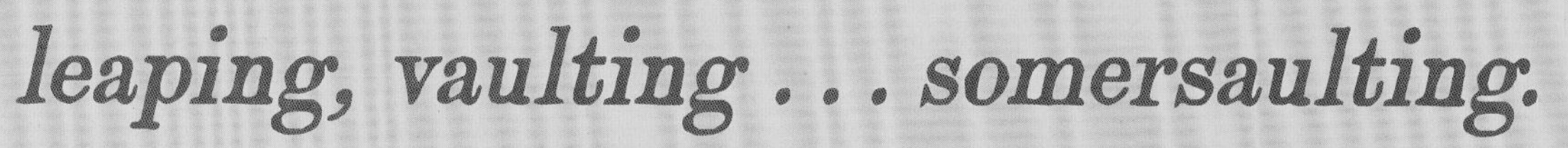

leaping, vaulting . . . somersaulting.

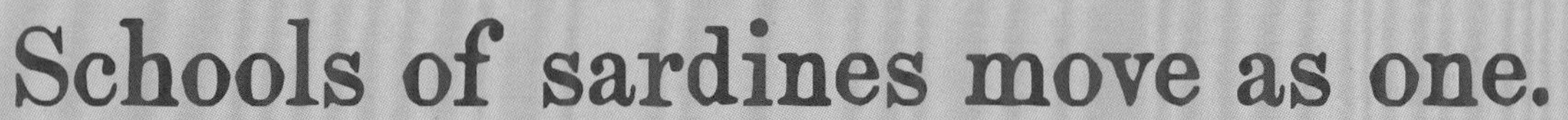
Schools of sardines move as one.

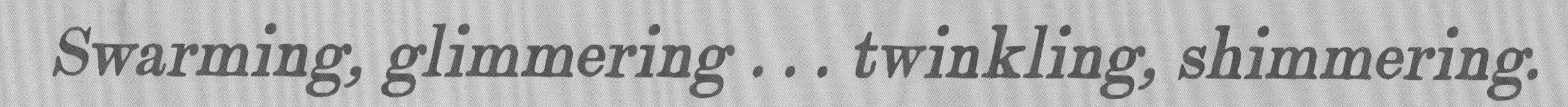

Swarming, glimmering . . . twinkling, shimmering.

Hungry hunters gather close.

 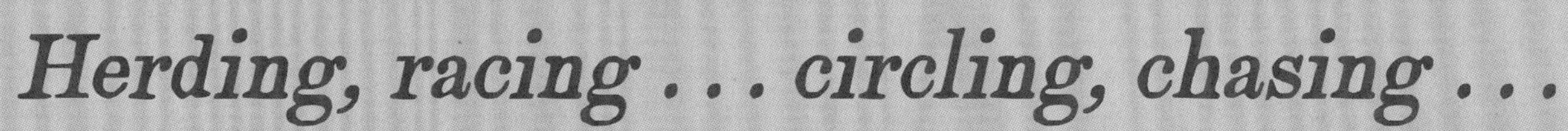

Herding, racing . . . circling, chasing . . .

Waters froth—a full attack!

Tails are thrashing . . . teeth are flashing!

Dive, dive! It's cold and dim.
One last stop, where sperm whales swim.

From rocky coast to deep down under,
the ocean wraps our world in wonder.

ONE OCEAN IN MOTION

The world's ocean covers more than 70 percent of our planet and provides homes for a vast diversity of living things, from tiny microbes to enormous whales. The ocean supports the water cycle and provides fresh air, food, and a habitable climate—everything that makes life possible on Earth. Its beauty inspires us to explore!

And did you know that there's only ONE ocean? Five major ocean basins—Pacific, Atlantic, Indian, Southern, and Arctic—make up a single interconnected system. Like the animals within it, that system is always in motion, powered by winds, tides, and Earth's rotation. A change in one area, such as pollution or overfishing, can impact marine ecosystems many miles away.

Learn more about the ocean and how to protect it:

NOAA 10 Ways to Help Our Ocean: OceanService.noaa.gov/ocean/help-our-ocean.html

Ocean Literacy: Marine-ed.org/ocean-literacy/overview#ocean-literacy-overview

Monterey Bay Aquarium Animals A-Z: MontereyBayAquarium.org/animals

For additional resources: ElizabethShreeve.com/educational-resources/

FIELD GUIDE

Most animals in this book inhabit the North Pacific. This vast area lies north of the equator and supports abundant wildlife, as in this scene from Alaska's Kodiak Island.

For centuries, Japanese free divers known as *ama*, or "sea women," collected shellfish, including the **abalone** pictured here. Abalone are also found along the California coast.

Sea otters keep coastal ecosystems healthy by feeding on **sea urchins** that destroy the seaweed known as kelp. The bright orange **garibaldi** is one of the kelp forest's showiest fish.

Intertidal zones, where ocean meets land, offer habitats for animals like **giant Pacific octopus**; the pink-and-red striped **flag rockfish**, and the round-headed **snailfish**.

Flying fish escape predators, such as the **mahi-mahi**, by leaping from the water and gliding on long, wing-like fins. **Red-tailed tropic birds** sometimes snag flying fish in midair.

Moonfish, or opah, have similar round shapes but average only 3 feet long. Ocean **sunfish**, or mola, lack the tails of most fish. At up to 12 feet long, they look like floating heads!

Black-footed albatross soar for hours on wings up to 7 feet long. The much smaller **Scripp's murrelets** also spend most of their lives at sea. The **rhinoceros auklet** sports a horn on its yellow beak during the summer.

Basking sharks can reach up to 40 feet in length, filter feeding on microscopic animals, called zooplankton. Another world traveler, the **blue whale**, reaches up to almost 100 feet in length.

Loggerhead turtles travel widely, nesting over the broadest range of any sea turtle. **Leatherback turtles** are known to swim 10,000 miles a year and dive to 4,000 feet below the surface.

Great white sharks are powerful, solitary hunters that reach speeds up to 35 mph when chasing prey, such as **elephant seals**. **Short-beaked common dolphins** are playful animals that travel together.

Pacific sardines roam waters off the west coast of North America in groups, known as "schools." These small fish provide food for many larger animals. Large schools of sardines can sometimes stir up bioluminescent plankton, creating a milky glow at night.

King salmon grow to 5 feet long, making them the largest Pacific salmon. Fast-swimming **skipjack** and **yellowfin tuna** share warm surface waters of the world with sharp-snouted **striped marlin**. All four fish are powerful predators.

Baitballs form when small fish, like **anchovies**, swarm together to defend against hunters such as **yellowtail tuna** and **wahoo**. Baitballs sometimes attract larger carnivores like **thresher sharks** and **spinner dolphins**. Thresher sharks stun their prey by thwacking them with their tails!

Sperm whales, the largest of all toothed predators, typically dive 2,000 feet down in search of food. **Humboldt squid** is one of their favorite snacks.

Hardhat divers once worked in the historic sardine fisheries of California, installing and repairing underwater equipment. While underwater, they would have spied kelp forest fish like this inquisitive **rock greenling**.

For Ray
—ES

For my grandnephews
Cameron and Easton, and the next
generation of ocean explorers
—RT

Printed in China by Dream Colour Printing Ltd.
in October 2024

LITTLE BIGFOOT with colophon is a registered trademark of Blue Star Press, LLC

29 28 27 26 25 9 8 7 6 5 4 3 2 1

Editor: Christy Cox
Production editor: Peggy Gannon
Designer: Tony Ong

Library of Congress Cataloging-in-Publication Data is available.

ISBN: 978-1-63217-539-7 HARDCOVER
ISBN: 978-1-63217-540-3 PAPERBACK

Sasquatch Books
1325 Fourth Avenue, Suite 1025
Seattle, WA 98101

SasquatchBooks.com